The Last Hours of Humanity

Warming the World to Extinction

Thom Hartmann

Waterfront Digital Press

TABLE OF CONTENTS

Chapter 1: The Extinction Threat

Imagine you and your family had a time machine large enough to live in no matter what era you arrived in.

Two hundred years ago there were only one billion people on planet Earth and farming was the major activity of not only most Americans, but most people around the world. A thousand years ago Europe was living in the Dark Ages, and the planet only had about 500 million people on it. Ten thousand years ago humans were just experimenting with agriculture and not even a single city had yet to be built in all of history.

50,000 years ago humanity had not begun to mine and refine metals. 100,000 years ago it appears we were just beginning to develop complex languages and societies and even further back to 200,000 years ago humanity had not yet emerged on the plains of Eastern Africa.

Modern humanity's entire history is less than a fifth of one million years. We have never lived through a massive asteroid strike to the Earth or other cataclysms large enough to produce a worldwide extinction of most life. But the planet has.

The worst was at the end of the Permian era, 251 million years ago. As I'll describe in more detail later, a massive lava flow in what is now Siberia initiated the process of warming the planet enough to destroy more than 95% of all life. Scientists call it the Permian Mass Extinction.

So let's imagine you're in your Time Machine dialed back to 251 million years ago, when you step out onto a fern-covered field somewhere thousands of miles away from what it is today, Siberia. Unbeknownst to you, the lava flow has begun in the distance.

As you're setting up your new home, you might notice a reddening of the sky at sunrise and sunset, but this wouldn't seem like anything dramatic was going on anywhere in the environment. Over the next few years and decades you may notice that the weather is growing more intense. For example the seasonal extremes are more noticeable. Still, you would have no way of knowing that the planet was moving toward a point that would lead to the death of almost everything.

Over the years, as you become an old man or woman, your children might notice that the larger plants, what today would be our trees, seem to be dying faster than normal from what appears to be some sort of blight or fungus. It's as if the plant's immune systems have been compromised.

Long after you are dead, your grandchildren might begin to notice that the rains are not coming the way they used to, and when they do, the storms are wildly more ferocious. The results are either floods or droughts. Insect and small animal populations are less evident. The air is becoming corrosive; thus it's getting harder to breathe. And the sunrises and sunsets are becoming more spectacular with radiant colors of light cast across the sky.

By the time your great great great great grandchildren are born, the lava flows may be settling down, but the tipping point has been passed. Although it was a slow process, it was an inexorable one. The world, over thousands of years, had warmed to the degree where all of the abundant life your descendants could see around themselves was doomed.

The lava flow in Siberia, by throwing into the atmosphere a variety of gases, had been steadily pushing up the temperature of the Earth. While many of these gases were themselves toxic, it would be their combined greenhouse effects along with those of the multiplying carbon dioxide levels that would take down the planet.

But you would only know that in retrospect when pretty much everything was dead.

Fast forward to today

The climate scientists' warnings have come true.

There is more carbon in the atmosphere trapping heat and moisture than ever before in the 165,000 years of human history. We are on the verge of the first ice-free summer in the Arctic in three million years — and back then the Earth was a much different place from the one currently cradling us.

The consequences of a warming planet are appearing much faster than had been projected by climate scientists of just a decade ago. The most dire warnings: rising oceans, freak storms and agricultural collapse, are all taking place now.

And it's going to get far worse.

But now, other voices have entered the fray. Those of geologists who study the longer term implications and histories of a planet undergoing rapid global warming. Specifically, they are focused on extinctions.

The climate scientists, geologists, and those from related scientific disciplines need to continue talking to each other because at some point we may be able to see the critical moment in which the current climate crisis crosses from the realm of a global destabilizer to a global extinction event. We must wake up.

A history of extinctions

It's hard to imagine Earth without life. We take the vast variety of life on this planet, and even our own existence, for granted.

But numerous times in our planet's history, life as we know it, has come close to disappearing entirely. We call these events mass extinctions and even teach schoolchildren about these times of great death on our planet. For example, we know that long ago on a much more unstable planet, the dinosaurs were killed by an asteroid striking the Earth. This leads many people to believe that as long as we don't see an asteroid hurtling toward the planet all is well.

But this is not rational thinking for several reasons. The asteroid impact that killed the dinosaurs and started a major mass extinction is the only event having something to do with outer space we can trace with any certainty. And new science indicates that the asteroid impact itself wasn't what killed the dinosaurs — it was the global warming that followed it.

New science has discovered a common theme in all of the extinctions in the past. And it's woven right into the global fabric of today as yet another mass extinction threatens our planet. That consistent thread is global warming.

There have been six mass extinctions in the billion-year history of life on our planet.

Each sharp spike in the chart below[1] indicates one of these mass extinctions.

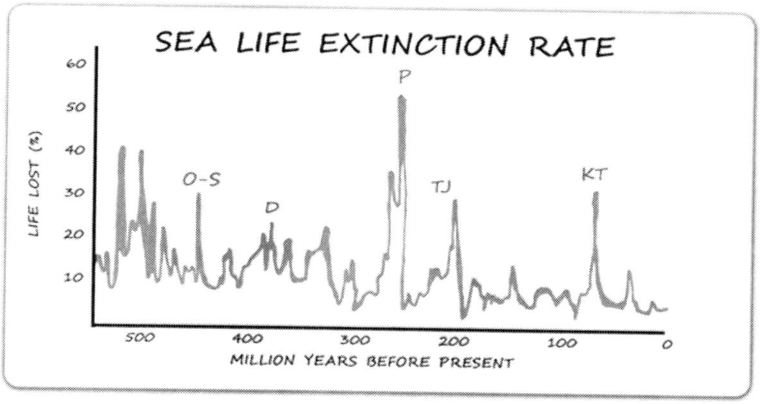

Occurring about 450 million years ago, the Ordovician-Silurian mass extinction devastated marine life, which at the time, dominated the planet. In a series of two extinctions, 60% to 70% of all life was taken from the planet.

Then, fewer than 100 million years later, the planet was rocked again. The Devonian period was capped off by a 20-million-year death march that also killed off 70% of life on the planet. This included many coral reefs which didn't return for another 100 million years.

We know of the Cretacious-Tertiary (K-T) extinction, which occurred 65 million years ago ending the reign of the dinosaurs. There was also an extinction event 200 million years ago known as the Triassic-Jurassic mass extinction.

But none of these extinctions explains the huge spike shown in the center of the previous chart. That one happened 250 million years ago, and was the worst mass extinction of species event in the history of our planet. It was the extinction of all extinctions. Referred to as "The Great Dying," the Permian Mass Extinction took out at least 95% of all life on the planet in less than 100,000 years — an instant in geological time.

Professor Paul Wignall with the University of Leeds and an expert on mass extinctions told me that the Permian was "the greatest crisis that life on Earth has ever suffered."

The true cause of extinctions

Only in the past two decades has the cause of the Permian extinction been understood. It was speculated that an asteroid impact may have been the trigger, but more recent research by Professor Wignall, geologists and other scientists around the world have revealed the true trigger came from deep within the Earth.

The Permian Mass Extinction was initiated by a colossal flow of lava in an area of what is now Siberia. That was the trigger but not the killer. The killer was under the water and under the ice where trillions of tons of greenhouse gases largely derived from carbon and frozen in the form of crystalline methane lay in wait.

Thus, global warming is the force behind the death of nearly everything on the planet during the Permian Mass Extinction. That point is illustrated in the chart below[2]. You can again see the spikes of mass extinctions measured by the increase in global temperatures, with the largest spike representing the Permian Mass Extinction.

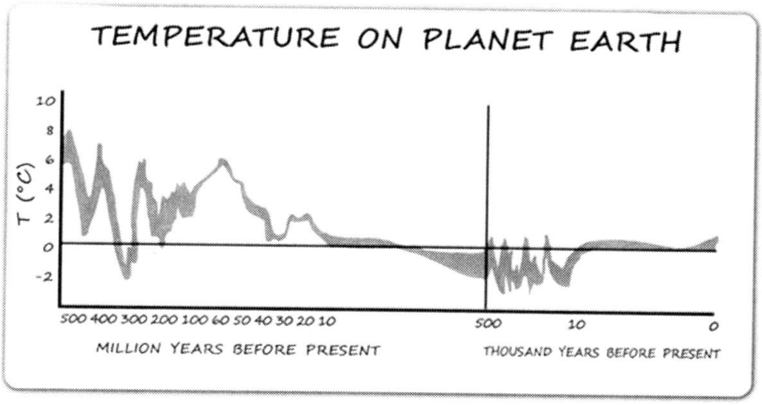

Wignall told me, "There've been a lot of disasters and crises in the geological past. It's interesting to study them because they may have a comparison to today."

He added, "I think it is certainly extremely significant that a lot of the main crises of the past are associated with global warming, with obvious implications for the present day."

The sixth mass extinction may even rival the speed and intensity of the great Permian Mass Extinction. But this sixth is not represented on either of the two previous charts. That's because it's the one happening today.

Right now.

All around us.

Chapter 2: The Great Dying… in fast forward

It was 1999 and my wife, Louise, and I were on the Big Island of Hawaii. Our host was an elderly sociologist, Dr. Robert Wolff[3], who had grown up in Malaysia. I first met Robert years earlier through his book "Original Wisdom," one of the most profound writings I have ever read. He wrote about his time and experiences in Malaysia living amongst the indigenous Sng'oi people.

Our host drove us to the edge of an active lava field in a caldera of the largest active volcano in the world: the Mauna Loa.

We hiked past the "Do Not Enter — Active Lava Field" signs and gingerly walked on a thin crust of recently cooled rock that was still warm to the touch. All around us steam and gases were slowly leaking into the air. Occasionally they spurted from holes in the ground that ranged from the size of a quarter to others measuring several feet across. In the distance was much more activity, red hot rocks and clouds of poisonous gases and steam. At the time I didn't realize it was a lava field like this, only different in size and intensity, which triggered the Permian Mass Extinction that killed off nearly all life on Earth in the distant past.

Robert guided us carefully as he talked to the spirits of the volcano. This place was sacred to him and to his relatives, local descendants of the aboriginal people of Hawaii. Then we performed a ceremony that I gave my word not to describe in writing.

The Siberian Connection

It's now frozen tundra but in the early years of the Permian Mass Extinction a part of Siberia known as the "Siberian Traps" was once an ocean of spewing lava. This is also where the "Great Dying" 250 million years ago started.

Michael Benton is a paleontologist at the University of Bristol. He dismisses the idea of an asteroid causing the Permian extinction. "People have looked for evidence of impact. But in this case, the evidence for impact is non-existent. But, there is great evidence for massive volcanic eruptions in what is now Siberia producing millions and millions of tons of carbon dioxide in the atmosphere. Carbon dioxide is a well-known greenhouse gas."

Rather than the typical, isolated volcanic eruptions we're used to, Siberia was torn up 250 million years ago by what's known as a "flood basalt eruption". That occurs when the crust of the Earth is split open and lava comes pouring out in great sheets. It's the largest type of volcanic eruption the planet generates.

Something similar, only on a much smaller level, happened not too long ago. Iceland experienced a flood basalt eruption at the end of the 18th century that spewed so much ash into the sky that Benjamin Franklin noted that the year 1784 came and went without a summer.

But the Icelandic eruption was nothing compared to the Siberian event 250 million years earlier. which scientists estimate lasted thousands of years and was characterized by a prolonged period of cooling. This cooling was from all that ash in the atmosphere, essentially creating a "nuclear winter" scenario. It was then followed by a rapid uptick in atmospheric greenhouse gases.

According to Vincent Courtillon with the University of Paris, it was a "volcanic eruption ten thousand times larger than man has ever seen." This cataclysm, however, was only the first stage of the extinction. As Professor Benton told me, "a string of consequences followed from that."

The effect of this massive volcanic eruption throwing so many greenhouse gasses, particularly carbon dioxide, into the atmosphere was at least a six-degree Celsius increase in global temperatures. However, despite having catastrophic consequences for most life on the planet, that six-degree increase in global temperatures could not do anything close to the damage to living species that we find in the fossil record during the Permian Mass Extinction.

Scientist Paul Wignall trekked to Iceland to examine Permian rocks for clues as to what caused that extinction. He deduced from fossil discoveries that the Permian extinction didn't kill off most life on the planet instantly but rather instead killed off land and sea life in three different stages over the course of roughly 80,000 years, a relatively short time geologically speaking.

He also found something else."Coincidental with the start of the crisis, we start to see a lot more carbon 12 appearing in the rocks, and that tells us something major is happening in the world's global carbon system."

Carbon 12 is the most abundant form (over 98 percent) of carbon on Earth: it has twelve protons, electrons and neutrons, and thus number 12 on the Periodic Table. It's formed in the core of stars through nuclear fusion, and bursts out into the universe when stars die and consequently explode.

Carbon 12 is also the root of all "carbon-based life forms" – the foundation of virtually all life on Earth. It's also a core ingredient of several greenhouse gases, most notably carbon dioxide (CO2) and methane (CH4), which, when released into the atmosphere, trap heat and warm the planet.

But carbon-12 doesn't come pouring out of the ground along with lava, at least not in the quantities Wignall found in the rocks that were being formed around the beginning of the Permian Mass Extinction. So, where'd it come from? The answer has to do with ancient plants and ancient sunlight.

Death by ancient sunlight

For over 160,000 years, the entire history of the human race, we have lived on current sunlight. Sunshine lit the Earth and drove photosynthesis. Plants absorbed carbon dioxide from the atmosphere and used photosynthesis and the sun's energy to incorporate that carbon into their roots and stems and leaves.

So we ate the plants, and we ate the animals that ate the plants. We burned the plants for heat and we used the plants to make our housing and clothing.

And then we discovered ancient plants, which had stored ancient sunlight. Plants — from plankton and algae all the way up to giant trees — have been absorbing carbon from our atmosphere and energy from the sun for hundreds of millions of years. Some of them, instead of decomposing immediately after they died, settled to the bottom of the sea or ended up buried under the Earth through geologic activity.

Over millions of years, those plants with their locked-up carbon and the energy from ancient sunlight turned into coal, oil and natural gas. We call these fossil fuels. Put a drop of gasoline in your hand or look at a piece of coal and you are looking at something hundreds of millions of years older than even most long-lived animal life on today's Earth.

On the ocean floors, principally around continental edges, there is also an enormous amount of carbon locked inside frozen methane. It is the result of millions and millions of years of plant matter dying in the oceans, settling to the seafloor, rotting and producing methane gas that becomes frozen into lattices of ice.

These massive beds of methane, which may exceed all other fossil fuels known to man, are stable for the moment because they're in deep water that is around 30 to 32°F. At those temperatures water also typically freezes, but in the depths pressure keeps the water liquid. That is unless it is saturated with methane at which point it then freezes into a snow-cone-like slurry called methane hydrate or methane clathrate.

Japan accomplished the first successful attempt to bring this methane hydrate to the surface in commercial quantities in 2013. It is their hope to add the compound to their portfolio of fossil fuels. Other than by drilling into it, methane clathrate crystals remain locked frozen in the sea beds for as long as the oceans are deep enough and cold enough.

But what happens when the waters suddenly warm? The methane melts, as Gerald Dickens of Rice University discovered. He was with a geologist friend at a bar in the late 1990's discussing how scientists were stumped by the presence of a burst of carbon-12 in some fossils remaining from the early stages of the Permian Mass Extinction.

"We were just sort of chatting about what we were working on for the summer," Dickens told the BBC and confirmed to me. "And he was telling me that he was working on trying to understand how you could get massive amounts of 12-carbon quickly, because some signatures had been found in the rock records and it was very difficult to explain, it didn't make any sense."

And then Dickens had an "a-ha" moment. He had previously worked for an offshore drilling company that was exploring the possibility of tapping into oceanic methane slurries and bringing the resulting natural gas to the surface. (As mentioned, the Japanese accomplished this in 2013). Dickens knew there were vast reserves of methane hydrate trapped frozen in sea-beds all around the world and wondered what would happen if a lot of that frozen methane on the sea floor had melted from a solid into a gas, and bubbled up from the ocean's depths? Would it be enough to account for the "signature" of carbon-12 that geologists were finding in the rocks associated with the Permian Mass Extinction?

So he went back to the lab and melted frozen methane in water warm enough. The results were dramatic. The gas not only dissolved into the water, but it also rose up out of the water and into the air. Dickens published a paper in 1999 suggesting that a 5 degree Celsius increase in ocean temperatures would have been adequate to melt enough methane hydrate crystals to create the Permian carbon-12 signature. And such a massive release of methane, itself a greenhouse gas vastly more potent than CO2, would also trigger a catastrophic, swift warming of the planet.

Permian explained

After reading Dickens' paper, back in the UK professor Paul Wignall projected that a worldwide melting of methane hydrate in the oceans would lead to a *second* five-or six-degree increase in global temperatures. The compounded warming suddenly made sense to Wignall. It was an incredibly elegant explanation for why 95% of everything alive on planet Earth was wiped out in such a relatively short period of time.

First, the colossal flood-basalt eruption in Siberia spiked global temperatures at least six degrees. That then warmed the oceans, which triggered the frozen methane hydrates trapped on the seafloor to melt. And, like soda pop, the highly potent greenhouse gas methane bubbled up through the ocean and into the atmosphere. That in turn would then easily have supplied enough methane to further rapidly increase temperatures on the planet to levels making Earth uninhabitable for most life-forms.

The BBC recently released a documentary about those dramatic new findings regarding the cause of the Permian Mass Extinction. They described the process:

> *…[T]here were two Permian killers. The Siberian Traps did erupt, contributing first to a nuclear winter cooling effect [caused by dust] and then to global warming [due to greenhouse gases]. Over 40,000 years, some land animals gradually died out, while life in the seas lived relatively calmly on, as the water temperature gently rose. Then the seas gave up their frozen methane. In just 5,000 years, there was massive loss of species from the world's oceans. In a third and final phase of the extinction, the Permian killer returned to stalk the land for another 35,000 years. By the end of that process, 95% of the Earth's species were extinct.*

Not an anomaly

A similar methane release happened during the K/T mass extinction, which killed the dinosaurs 65 millions years ago. Only instead of massive volcanic eruptions in Siberia, it was an asteroid splashing into the Gulf of Mexico.

Peter Ward, a paleontologist at the University of Washington, Seattle, explained to me, "We now know that the dinosaur-killing K/T mass extinction has within it a carbon dioxide aspect as well." He added, "We're seeing a unification, that mass extinctions are caused by increases in greenhouse gasses."

About ten millions years later, there was another rapid warming of the planet known as the Paleocene-Eocene Thermal Maximum. It's unknown what the trigger for it was, but global temperatures increased by six degrees Celsius in as little as 20,000 years, more than enough to trigger the melting of the oceans' frozen methane and the extinction tipping points behind the Permian and K/T events. Though it's not considered a mass extinction, the PETM killed off at least 35% of all life and even more in the sea where presumably the methane release turned the sea-waters toxic.

So we now know the formula for extinction. Something happens to increase global temperatures five to six degrees, which triggers a melting of the frozen carbon and methane oceanic reserves that then leads to further global warming devastating life on Earth.

Thus, the pressing question for us today is this: *Can seven billion people on the planet burning fossil fuels imitate the sort of carbon greenhouse gas release caused by the Permian lava flows, or the K/T mass extinction impact or whatever warming caused the PETM?*

The answer is yes.

Professor Wignall confirmed to me, "The numbers are very similar from some of these giant lava flows in Siberia. The amount of carbon dioxide they released was very similar to the fossil fuel burning, carbon dioxide release that we're doing decade after decade today."

Benjamin Black at MIT has been studying the carbon releases during the Permian. He told me, "It's kind of a scary thought, but one of the best geological analogues for the kinds of rapid changes in climate and CO2 in the atmosphere that we're going to witness now and for the next few centuries potentially is this end Permian time 250 million years ago that culminated in one of the largest mass extinctions that we know of."

And Dr. Michael Benton pointed out, "Looking at these ancient events shows us times of global warming, and the atmosphere doesn't care whether the carbon dioxide comes from human activity or a volcano. It has the same end effect."

Only this time, it could happen much faster than any of the previous extinction events. For example, in the Permian-Eocene Thermal Maximum (PETM) and the Permian Mass Extinction it took 10,000 to 80,000 years to warm the Earth six or more degrees. In comparison, we are burning ancient carbon-based fossil fuels so vigorously that we may soon have warmed the Earth six or more degrees in just a few short centuries.

Chapter 3: The Methane Overkill

When the mathematics and physics become as compelling as they are now, big business also gets alarmed. In November of 2012, the international accounting firm Price, Waterhouse, Cooper issued a rather dire warning for the future of business on the planet. After all, if the human race has been reduced to living in caves, or is extinct, nobody is going to buy or sell anything.

In their news release[4] sent out internationally and entitled "*'Business as Usual is Not an Option' as current rates of emissions reduction point to 6C of warming*," Jonathan Grant, director of sustainability and climate change, said:

> "The new reality is a much more challenging future in terms of planning, financing and predictability. Even doubling our current annual rates of decarbonisation globally every year to 2050, would still lead to 6C, making governments' ambitions to limit warming to 2C appear highly unrealistic."

Just a few months after the Price, Waterhouse, Cooper report, the World Bank came out with a detailed study confirming that unless radical actions are taken soon, the world could well be looking at a 5 degree Celsius increase in temperatures, purely from our burning of fossil fuels. Their executive summary included this chart below[5] — the black line shows the IPCC's (Intergovernmental Panel on Climate

Change) 1999 consensus of virtually all the world's major climate scientists' models of what will happen if nothing of consequence is done:

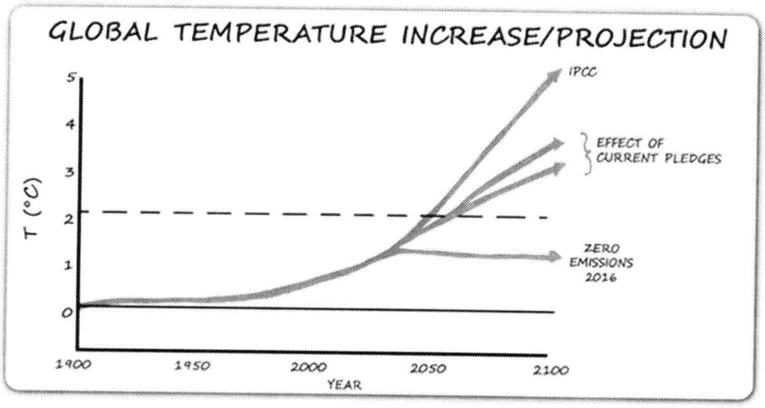

Global warming has been at the center of political and media debates[6] for a generation or two. While each year we see its devastating effects with more clarity, one consistency in the discussion has been the *underestimation* of how quickly the planet is actually warming, and how quickly sea ice in the Arctic is melting.

What these latest estimates don't point out is that a 6 degree Celsius increase in global temperatures has happened only once before in our planet's history — during the early, set-up years of the Permian Mass Extinction. While the Permian Mass Extinction almost certainly took the planet up to 12 degrees later, the entire PETM event appears to have been triggered and sustained by a mere 6 degrees Celsius. Virtually every animal larger than a dog that existed at the beginning of the PETM was dead or had evolved into a different life form by the end of the PETM. This is the *real* "inconvenient truth" that many are talking about amongst themselves, but few have been willing to say it out loud.

It's been a decade since scientists have suggested that the Permian Mass Extinction was caused by the one-two punch of a massive volcanic event that warmed the Earth, and this warming melted and released methane crystals from the oceans to warm the planet's atmosphere enough to kill off most life.

During that time, we've gone from a debate over whether our increasing levels of CO2 are warming the planet at all, to documenting a .8 degree Celsius increase just over the past three generations.

Researchers at Oregon State University and Harvard put out a study showing just how radically fast world temperatures are increasing. Published in *Science* in March of 2013, that study found that temperatures are changing *50-times faster* than they did at the onset of modern civilization and agriculture.

Coming off the hottest decade of global temperatures ever recorded, the study concluded that the planet is hotter today than it's been during nearly 80% of the last 11,300 years. They found that in the last 5,000 years, the planet cooled about 1.3-degrees (a slow, natural and accountable variation), but in just the geologic eye-blink of the last 100 years it's warmed 1.3 degrees — an amount and speed which no natural process could explain.

Another flood basalt eruption on the scale of the Siberian one that kicked off the Permian Mass Extinction isn't occurring, and an asteroid from outer space hasn't hit us.

However, humankind, with our obsession with burning fossil fuels, are mimicking both.

And we now know that this rapid warming caused by CO2 and CH4 emission is almost certainly just the first in a three-act play that could eventually kill off most life on Earth. It looks like the second act will be the Arctic giving up its methane.

The bomb under the ice

Since 2011, a team of NASA scientists has been flying a souped-up C-23 Sherpa aircraft low across the Arctic. The pilots describe their plane as a "UPS truck with a bad nose job." What they're finding suggests that rapid changes to the carbon cycle, which some believe are similar to what caused the Permian Mass Extinction and the PETM, are again underway.

Through the Carbon in Arctic Reservoirs Vulnerability Experiment, or CARVE, NASA is leading a five-year campaign to study how climate change is affecting the Arctic's carbon cycle. The scientists are looking at the vast reserves of frozen carbons locked in the Arctic, measuring how quickly they're melting and determining what sort of greenhouse gases the melting is releasing such as carbon dioxide or the much more toxic methane.

As Charles Miller, lead researcher on the CARVE team told me, "The climate models that we have for the Earth suggest climate change will happen to the largest extent in the poles, especially in the Arctic over land." He added, "The Arctic is the canary in the coal mine for the rest of the Earth system."

Over the last thirty years, temperatures within these permafrost soils have increased by 1.5 to 2.5 degrees Celsius — faster than even Arctic air temperatures which are increasing more rapidly than anywhere else on the planet. So far, the results of the CARVE experiments have been startling. To begin with, there's a mind-boggling amount of carbon locked up in the Arctic with the potential for release into the atmosphere.

During the previous extinction events, it was the large reservoirs of frozen carbon and methane buried underneath the oceans that, once they began melting, triggered the critical phase of the mass extinctions. In the millions of years since the last major warming period, the world's oceans and the Arctic have replenished their buried carbon reserves. And today they're melting again.

A lot of the Arctic carbon and methane deposits are found in permafrost zones which make up roughly one quarter of the Northern hemisphere's land mass.

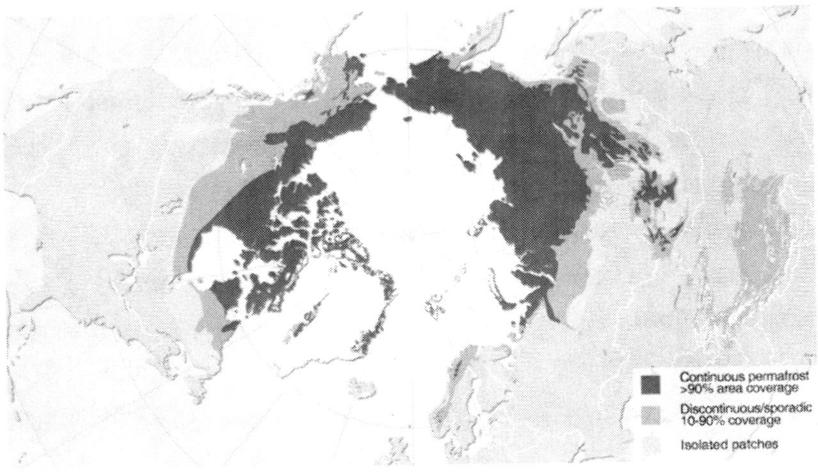

The map above[7] doesn't do justice to just how much carbon is stored in these permafrost zones.

Miller told me, "It's thought there are approximately 1,800 billion metric tons of carbon sequestered in the frozen soils and peat lands and permafrost in the Northern Atlantic and northern high latitudes."

To put that in perspective Miller explained, "That's a huge amount of carbon that could provide a natural feedback, compared to the 350 billion metric tons that have been put into the atmosphere by anthropogenic fossil fuel combustion since the beginning of the Industrial Revolution." We've burned a heck of a lot of fossil fuels since the Industrial Revolution but what Miller is saying is that in the Arctic alone, the carbon reserves in just permafrost and peat lands could increase global carbon emissions by more than 500%.[8]

The timeline according to Miller goes something like this, "By 2040, as much as 20 to 50 billion metric tons of the carbon that's currently frozen in the permafrost soils might become available. By 2100, those estimates range from 160 to 250 billion metric tons. So that by the end of the century you might have perturbation that would be almost as much as all the fossil fuel carbon that's been released since 1850."

Miller warned, "If even a small fraction of the 1,800 billion metric tons of carbon that is currently captured in those soils was to be released, it would have a significant impact on the greenhouse gas forces not only on the entire Earth system but also especially with concentration in high northern latitudes. That could then provide additional warming which could lead to mobilization of more of this carbon and something that would begin spinning out of control."

The biggest question is what percentage of this carbon release will come in the form of methane, the most potent widely available greenhouse gas?

When the permafrost melts, it releases either carbon dioxide or methane depending on the climactic changes occurring around the permafrost. A warmer, wetter climate will produce carbon dioxide emissions, while a warmer, dryer climate will produce methane emissions. Over a twenty year time period, methane is 105-times more potent a greenhouse gas than carbon dioxide, so it's critical to determine just what exactly is being released in the Arctic.

Miller said the CARVE team recently found,, "Some of the methane and carbon dioxide concentrations we've measured have been large, and we're seeing very different patterns from what models suggest." One example Miller cites is a reading the team got over the swamps in the Innoko Wilderness. There were methane levels 650 parts-per-billion higher than what's found in "normal backgrounds." "That's similar to what you might find in a large city," Miller added.

As the permafrost continues to melt, if just 1% of its carbon is released as methane, it will have the same potency as the 99% released as carbon dioxide. Currently CARVE researchers think the fraction of methane-to-carbon dioxide release is 3%. But Miller cautioned, "Pure methane release from the oceans on any large scale would be a significant greenhouse gas forcing. And even a small change in this 3% fraction of methane to carbon dioxide that we believe is being released by permafrost thaw, changing it even by 3 to 4%, would also have a significant impact on the greenhouse gas forcing."

But CARVE is just looking at the carbon and methane frozen in the Arctic permafrost and peat lands. The scary truth is that there's also huge potential for methane releases in the deeper oceans of more temperate latitudes.

Chapter 4: The "Clathrate Gun"

The deepest place on Earth, the deepest part of our oceans, is a massive canyon in the Western Pacific called the Mariana Trench. More than 35,000 feet deep, the pressure of the water is 1,000 times that of the pressure at the surface. At that depth, a normal submarine would collapse like a paper bag. There's very little life there, but what does live is adapted to cold, pressure and darkness like nowhere else on Earth.

The first man to dive solo to the bottom of the Mariana trench was filmmaker James Cameron. A lay expert on the oceans, in part from his work making the movie *Titanic*, Cameron designed, funded, built, and, in 2012, navigated his own submarine, the *Deepsea Challenger*, into those icy depths.

To say that James Cameron understands the oceans in a way that few other human beings ever could would be an understatement. From shallow dives to the deepest place on Earth, he knows the seas. And he knows about their methane clathrates. As he told me earlier this year, "The bottom line is that, given the number of species expected to die off in the next century just due to climate change and the other effects of our industrial civilization, we are ALREADY in the middle of a mass extinction event equivalent to the lesser ones in paleo-history, such as the K-T Extinction that took out the dinosaurs. Though the homo sapien isn't a species that's currently endangered.

"However the clathrate gun scenario changes that. It doubles down on an already bad situation, putting us on the hit list as well. Not literally, because some people will survive, in highly artificial environments meaning the species will persist but the vast majority of humanity will perish."

Cameron added, with a metaphor worthy of his brilliant *Avatar* movies, "We're talking about the survival of our kids. We, collectively are playing Russian roulette. We're putting a gun to their heads and going click, click, click. A clathrate gun."

Torches in the seas

Only discovered in a serious way in the 1960s, there is so much methane in the form of clathrates, such as the frozen methane slurry on the floor of the continental shelves around the oceans of the world, that it's at least double all the known underground coal, oil and gas on the entire planet.

Peter Ward, the paleonotologist at the University of Bristol, told me, "Methane is way worse than carbon dioxide. It's inert right now in the soil. It's not affecting anyone in any way. When you warm it though, it becomes a gas. Then it starts acting immediately as a greenhouse gas. So this is an immediate and very short term threat to planetary civilization."

For the last decade, a joint U.S.-Russia research team out of the International Arctic Research Centre at the University of Alaska, Fairbanks has been exploring the East Siberian Arctic. They are looking at the bottom of the sea-floor where once stable reserves of frozen methane have begun thawing and bubbling up into the atmosphere. Dr. Igor Semiletov led the research team in 2011. He told the British newspaper, *The*

Independent, about what he found that year and the scope of the impending methane problem. "Earlier we found torch-like structures like this but they were only tens of metres in diameter. This is the first time that we've found continuous, powerful and impressive seeping structures more than 1,000 meters in diameter. It's amazing."

Amazing, academically that is. Horrifying for those of us who call planet Earth home. A thousand meters is a kilometer so this one single massive methane leak is a bit more than a half-mile in diameter. And, it is just one of many.

Dr. Semiletov added, "I was most impressed by the sheer scale and the high density of the plumes. Over a relatively small area we found more than 100, but over a wider area there should be thousands of them." These sights would have been familiar during the Permian and other extinctions as well.

As Paul Wignall explained to me, "The significant thing about methane is it's a very effective greenhouse gas, so it causes global warming. Once you trigger warming, if you release methane, it's in positive feedback. It gets warmer and warmer as you release more methane into the atmosphere. So, we think this was a big part of the [Permian] extinction at that time — this very rapid warming."

Charles Dickens, who had proposed the methane link to the Permian Mass Extinction, told me, "We know that in the bottom of the sea floor, large parts of the ocean margins have methane in a solid phase. And what happens is when you change the temperature, it can melt this frozen methane phase." He added, "The idea is that during some of [these extinction] events we have some triggering that forces the ocean temperatures to warm and then that melts the solid methane phase which turns into gas and goes into the oceans and atmosphere."

Paleontologist Michael Benton concluded, "The risk is runaway greenhouse, which is when the self-correcting mechanisms cease to kick in and you heat by a little bit and you release methane. That then causes excess heating and you release more methane, and so it goes on… It's believed that at the end of the Permian 250 million years ago this kind of runaway greenhouse exacerbated the effect making it [the Permian Mass Extinction] an awful lot worse."

By 2010, it was said that 8 million tons of methane emissions were being released into the atmosphere every year, but researches believe this too is an underestimate. More recent studies have shown a far more rapid increase in methane (CH4) concentrations in the atmosphere. The chart below illustrates this trend. In the last two centuries, the amount of methane in the atmosphere has increased from .7ppm to 1.7ppm. That's the largest known increase in the history of the planet.

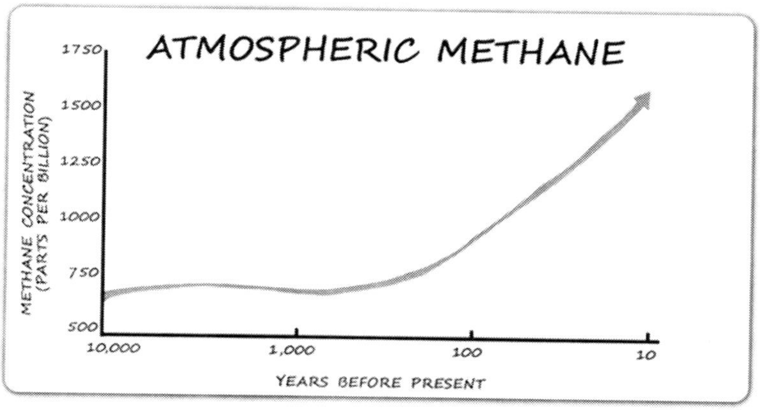

Russian biogeochemist Natalia Shakhova calculates that when the Siberian permafrost currently sealed into soils and underwater around the edges of the Arctic Ocean completely melts, methane concentrations in the atmosphere will increase twelve-fold.

Such a methane increase could, in a matter of just a few years, accelerate the warming of our planet beyond four to five degrees Celsius and into the extinction range of eight to ten degrees.

This doomsday scenario is called the "Clathrate Gun Hypothesis." It suggests that climate change will happen much quicker than most projections, since most projections fail to take into account the possibility of a sudden release of methane from the oceans when global temperatures rise five or more degrees like they did during the Permian, during other mass extinction events and the PETM.

Climate scientist Michael Mann was responsible for calculating the startling "Hockey Stick Chart"[9] below illustrating the rapidity with which global temperatures are increasing since man began burning fossil fuels.

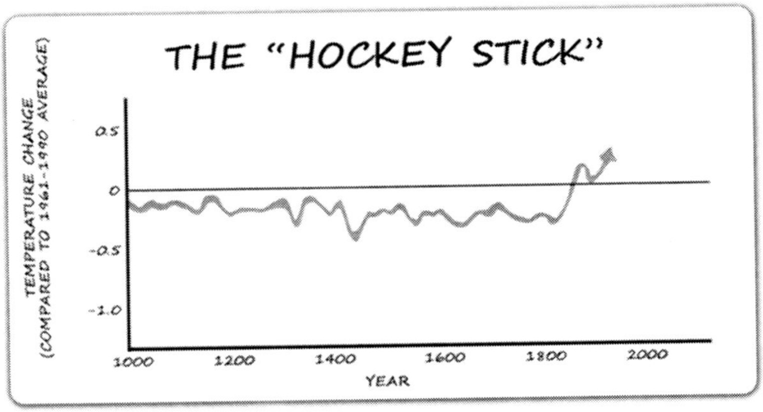

Looking at how quickly temperatures are increasing now, Mann told me, "We are literally releasing the carbon dioxide that nature had locked up over a hundred million [years] down below the Earth. And we're releasing all that carbon dioxide now at a rate a million times faster [than it accumulated]. It's the time frame of a century, not 100 million years."

Given the speed at which warming is happening, it's difficult to even compare today's events to the Permian and PETM. As Mann noted, "There is no evidence when we look to the past for any precedent for the rate of change in atmospheric composition that we're causing, and the rates of change in climate that we can expect, as we continue to burn fossil fuels and elevate these greenhouse gas concentrations."

Yes, more than seven billion humans burning fossil fuels *can* and *are* mimicking the forces that led to previous extinction events on the planet — from the Permian to PETM. So the question now shifts. Is there anything that can be done to stave off extinction at this point, or is it too late?

Chapter 5: Tipping Toward Extinction

Back in 1996, I noticed a recurring theme in news stories about oil. The Saudi's had recently restated by almost a third the amount of oil they had in the ground and other nations around the world were doing the same.

At the same time, geologists and oil scientists were pointing out for the last 10%, 20%, and in some cases as much as 50% of the oil in a well was so expensive to suck out from the ground that the extraction process cost more than the oil was worth. That means this last bit of oil would likely be abandoned and written off unless the price of oil increased dramatically to make it fiscally worthwhile for the oil companies.

The recurring theme amongst these geologists and scientists was something called "peak oil" and it was calculated as the point at which the maximum available amount of oil on Earth had been reached. Back then oil was selling for just a bit over $20 a barrel although there had been quite a spike more than a decade earlier during the Arab oil embargoes.

The predictions were that once peak oil was hit, the price of oil would rise substantially, at least double or triple, and thereafter would always be more expensive.

I wrote a book then entitled "*The Last Hours Of Ancient Sunlight*," which started with the premise that we were approaching peak oil, moved into the stupidity of building a civilization on fossils, and then raised a series of questions and hopefully answers about the impact of all this on our environment and our society.

Peak oil, I suggested, was a tipping point. Before it, oil is abundant and anxiety-free. After hitting peak oil though, the fuel would be expensive and nations — particularly the very oil-dependent United States — would go to war in the Middle East to get more of it. Peak oil was part danger, part opportunity. But either way, life would never be the same as it was before the peak.

The tipping point to extinction

Our planet is resilient. It's good at maintaining stable temperature levels by cycling greenhouse gasses containing carbon to keep the atmosphere suitable for life. But as carbon in the atmosphere reaches too high a level, these self-correcting mechanisms break down. This is how it has happened in every previous mass extinction event in the past.

Professor Michael Benton explained, "There's something we have to be aware of. The Earth's systems can correct for an excess of carbon dioxide and for other changes… But, from time to time, if you build carbon dioxide to a sufficient level, these normal feedback processes won't pick up and won't be sufficient. They won't be able to cope. Then you get to the so-called tipping point when everything falls apart."

Tipping points exist all around us. A glass of wine sitting on a table is in a stable state. Apply pressure to one side of the glass and you begin to alter that state. Eventually, a tipping point is reached where the state of the glass rapidly transitions from being upright, stable, and full to horizontal, stable and empty. In the middle, as it fell, was a point of great instability.

A tipping point is the moment in time when a process has exceeded its own ability to remain stable and stop itself from rapidly transitioning into a new stable state. Like an avalanche triggered into a tipping state by something as innocuous as a loud noise or the small additional weight of one, final snowfall, once a tipping point has been reached it must play itself out until a new, stable state is achieved. When the avalanche finally hits the bottom of the mountain, it stops moving, having achieved a new stable state.

When a mass extinction begins rolling down the hill of time, the geologic record shows that it picks up speed rapidly and then ends with a mass die-off of life. That mass die-off exhausts the processes that drive change in the climate and things settle down for a while. Then, slowly, life claws it's way back from the edge. For example, if climate change were to hit a tipping point where it destroyed human civilization, or even killed off the human race, it would stop us from pumping ancient carbon into the atmosphere and the atmosphere would achieve a new, steady state.

This is not however the best way to stop global warming. But, on our current course it is a very real possibility. The signs of extinction are already all around us. We are hitting or close to hitting a variety of tipping points.

Accelerating

A team of researchers at the University of Alberta recently published a paper in the journal *Global Environmental Change* characterizing the dangerous practice of low-balling estimates of global warming: "[T]he available evidence suggests that scientists have in fact been conservative in their projections of the impact of climate change…particularly in IPCC assessments of the physical science." They go on to say, "[S]cientists are not biased toward alarmism but rather the reverse: toward cautious estimates…We call this tendency 'erring on the side of least drama.'"

And they point out that while "erring on the side of least drama" may somewhat reduce the probability of big dollar, high-profile attacks from oil industry and billionaire funded pundits and think tanks, it does not do a service to science or society.

The facts bear this argument out. Below is a chart[10] showing IPCC projections for sea level changes around the planet. Notice how the red trend line (showing actual sea level increases) rises at a faster clip than the blue shaded area (which includes the IPCC's 2001-2007 projections from 1990-1999).

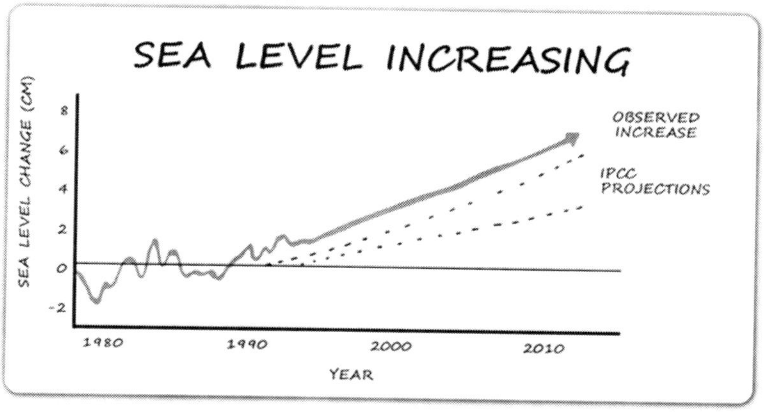

The same is true of IPCC projections of sea ice loss in the Arctic shown below[11]:

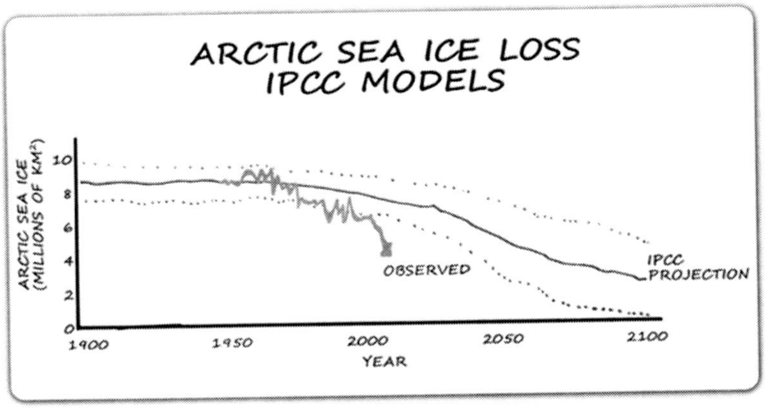

These miscalculations in the Arctic are especially troubling because what's happening there is critical to understanding what will happen to the rest of the planet as global temperatures reach levels not seen since the last "Great Dying." Even more critically, the Arctic is the edge of massive stores of permafrost which have trapped billions of tons of carbon that will be released as methane and carbon dioxide as the permafrost thaws.

It's likely that a tipping point-of-no-return has already been reached when it comes to the Arctic sea ice, which doesn't bode well for efforts to keep the methane frozen in the oceans. According to climate scientists Michael Mann, ice in the Arctic may already be a lost cause: "With the ice sheets, it's possible that we've already warmed the climate enough that the Greenland ice sheet will continue to melt."

He told me, "Even if we reduce CO2 emissions, even if we bring CO2 concentrations to peak, it's possible that the Greenland ice sheet will continue to melt. It's possible that the Antarctic ice sheet will continue to melt."

It's estimated that the Arctic, within seven years and maybe as soon as 2015, will have its first ice-free summer in the last 700,000 years (keep in mind that humans have only been on this planet for 165,000 years). Earlier projections predicted ice-free summers as far out in the future as 2080.

Losing all that ice would further accelerate global warming. Unlike the ice, which reflects heat, blue open waters absorb heat. The larger the area of ice-free waters, the more heat is absorbed into the seas accelerating the rise of ocean temperatures and eventually unlocking the methane clathrates.

One explanation for these projection-misses, beyond politics which we'll explore later, is that scientists have also been consistently underestimating how quickly CO2 emissions would increase, as the chart below[12] shows:

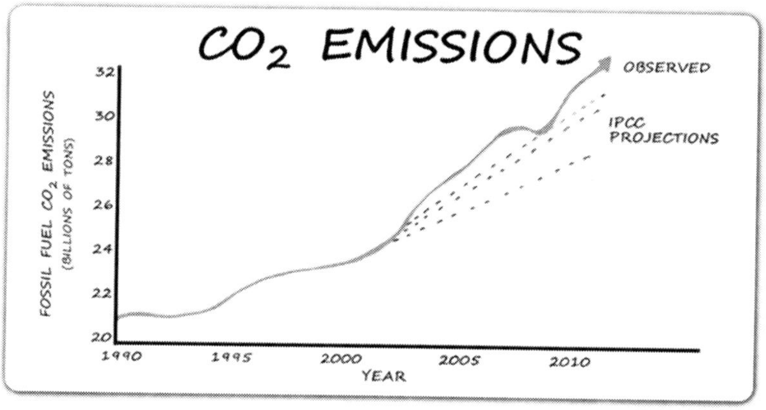

The atmosphere currently holds about five percent more water vapor and carbon dioxide (CO_2) than it did for most of the rest of the time humans have been on this Earth. And with those changes in atmospheric CO_2 and moisture have come increased climatic and oceanic instability. Bad storms are worse than historical norms, sea levels are beginning to creep upwards, and most importantly, the oceans are becoming warmer and more acidic.

Mann noted, "It's possible that the acidification of the oceans because of the increasing CO_2 levels in the atmosphere…will sort of push these ecosystems off the edge and we could see a rapid decline. Some scientists fear that we could see that now in a matter of a decade or two. We could see the disappearance of the Great Barrier Reef in Australia, for example."

A 2011 report by a panel of leading marine scientists working with Oxford University, the International Programme on the State of the Ocean (IPSO) and the International Union for the Conservation of Nature (IUCN) called the current extinction of life in our oceans "unprecedented in human history." The panel noted that the speed at which marine life is dying off increased beyond projections and that an extinction event rivaling even the Permian extinction may already be underway.

Beginning in the 1970s, predictions about the consequences of global warming have mostly been couched in the language of storm damages, ocean rises and the dangers of changing land and sea ecosystems. The main message overall has been that things are getting bad and will get worse.

The most dire warnings from people like NASA's James Hansen suggest that it could even get difficult for humans to survive on a planet with a climate substantially different from the one on which we evolved.

In a 2008 paper[13] co-authored by Hansen, he noted that, "If humanity wishes to preserve a planet similar to that on which civilization developed and to which life on Earth is adapted, paleoclimate evidence and ongoing climate change suggest that CO2 will need to be reduced from its current 385 ppm to at most 350 ppm."

Five years later in May of 2013, the Mauna Loa Observatory in Hawaii recorded that the carbon dioxide concentration in our atmosphere had reached 400 ppm. Hansen's line in the sand wasn't just crossed, it was obliterated at a faster rate than any of the world's high-profile scientists had predicted.

I asked Michael Mann about this and he answered, "The problem with these tipping points is we don't know exactly where they lie. We know that they're out there. But, where we cross those tipping points, we don't know for sure."

That, in fact, is one of the biggest problems with tipping points. They are almost always seen in retrospect. By the time you're crossing them it's often impossible to know exactly what is happening. Thus, the perennial habit of underestimating greenhouse gas emissions and global warming combined with unknown tipping points is a dangerous mix.

Our best guesses

The international consensus has pegged a tipping point at two-degrees Celsius meaning any temperature increase above must be prevented at all costs. Already humans have increased global temperatures by nearly half that, .8 degrees Celsius, in just the past few decades. And we've seen what just that small increase has done to our planet. This is an incredibly daunting task we face given that current models are calculating an increase of five to six degrees within the next 100 years.

Michael Benton told me, "The consequences are there. We know that. You can't do something without having a consequence. More and more people are seeing the climate changing, sea levels rising… there is hard evidence that temperatures have risen by about 1-degree in the last 50 [to] 100 years… It's east to rationalize it away. But, then another degree, another degree and then you do get toward the tipping point."

It's also important to remember that because of positive feedback loops associated with global warming, a .8-degree Celsius increase in temperatures means that a full 1.5-degree increase is inevitable due to the compounded effects of melting ice, ocean acidification and methane releases. The net effect of these cumulative, positive-feedback reinforcements is that whatever amount of warming we have seen from a given point in time will double in the future. Even if all human-produced carbon stops flowing into the environment immediately.

So, considering the warming that's taken place we're already close to the two-degree cap. And if nothing is done before we hit a 1.5 degree increase, which is likely within the next 20 to 30 years, then a 3-degree increase will be inevitable, and so on.

The man who helped put global warming on the map, climate activist Bill McKibben, has thrown out a new number that could signify another potential tipping point: 565 gigatons. That's the threshold for the amount of carbon dioxide we can release into the atmosphere to keep temperature warming less than two-degrees.

But, here's the problem. The amount of fossil fuel in current reserves and expected to be burned by oil companies and oil producing countries would lead to a release of 2,795 gigatons — more than five times greater than McKibben's tipping point theory.

In 2011, the world put out 31.6 gigatons of carbon. If those numbers stay steady, though they're currently increasing, then we'll hit McKibben's threshold within two decades. Using his math, that's how much time we have left. Beyond that, any effort to save the planet may be fruitless.

Chapter 6: Staving off Extinction

In 1996 my wife, Louise, and I sold an advertising agency we had started in Atlanta to our employees, and moved to Vermont. Determined to dial back our fuel usage (I was then researching *The Last Hours Of Ancient Sunlight*), we bought a Honda Insight hybrid car. The vehicle was incredible. Several times over the next two years we drove back to Michigan, as my father had a couple of heart attacks, and to visit family for holidays. I had gotten almost excessive about tracking the gas mileage and had it up to 73 miles per gallon on a fairly regular basis.

Ultimately though my back took priority. I had broken it skydiving around the time I turned 20, and there just wasn't room to sit up straight in the Honda Insight. So we traded it in for a Prius.

The thing that the Insight put me most in touch with was how rarely I drove long distances. I would fill the tank only once a month or so, and the tank only held ten gallons. Turns out I wasn't unique in that. Most Americans — 78% to be exact — drive fewer than 40 miles per day. Which means that for more than three-quarters of us, we really don't need gasoline at all. Electric cars have an easy minimum range of 100 miles and a typical range of closer to 200 miles. And the technology is just getting started.

We fight wars all over the world for oil, and we don't even need it. That, in and of itself, is insane.

But when you look at a 250 million-year-old piece of sedimentary rock and see that it's virtually barren of fossilized life forms, you realize this is a hell of a lot more important than whether or not our country is going to fight stupid wars in far-off lands so the oil companies and the Koch brothers can continue to get rich. This is about the survival of life on this planet, at least as we know it. To hold back extinction, the human race must immediately limit carbon emissions.

There are a number of ways to do this. To start with, governments around the world can simply set a limit on carbon emissions and begin fining businesses that pollute beyond that cap.

If we want to make it more politically palatable and throw in some profit for the "banksters" who own so many of our politicians, we could also let businesses and banks set up an exchange and trade carbon credits. Businesses can buy the right to pollute, or sell the consequences of their good behavior. In that way the "banksters" get to skim a piece off the top just like they've managed to do with pretty much everything else.

This is our best chance to stave off extinction. However, it's politically impossible at least for now. Although as the consequences of global warming further manifest themselves, political will for a hard cap will undoubtedly build, just like it did with sulfur dioxide that caused acid rain in the 1980s. It's rarely discussed in the press, but President George Herbert

Walker Bush successfully pushed through a cap-and-trade program for sulfur dioxide which radically reduced acid rain.

To buy humanity time in the short-term we may have to settle for a more modest proposal such as a carbon tax. The concept is simple: put a price on pollution. Under a carbon tax system, a business would pay, for example, $20 for every metric ton of carbon they spew into the air.

As of 2013, 33 countries and 18 smaller jurisdictions around the world have some sort of carbon tax in place. In February of 2013, China, the world's largest polluter of carbon, announced plans to implement a carbon tax. And the European Union runs a carbon cap-and-trade system.

In 2010, Congress nearly passed a comprehensive cap-and-trade law that would have kept carbon emissions at a rational level and required businesses that wanted to pollute beyond the limit to buy carbon credits on a market from other businesses that were polluting below the level. Though that measure died by a Republican filibuster in the Senate, elsewhere around the world carbon taxes are flourishing.

Alongside the EU's cap-and-trade system, several other European nations are taxing carbon. They include the nations of Scandinavia, the UK, the Netherlands, Switzerland and Ireland.

Other like-minded countries include Costa Rica, which funnels tax revenues from carbon polluters to indigenous communities to help fight deforestation; South Africa, where new auto sales include a carbon tax and full-scale carbon tax across most other industries has been implemented; and India, which has a carbon tax of 50 rupees per ton on coal produced or imported into India.

South Korea and Japan have also passed carbon tax measures. Several provinces in Canada have carbon taxes. And here in the United States, there are local carbon taxes in San Francisco, Boulder and Montgomery County, Maryland while quietly, business leaders and even Republican lawmakers are coming around to the idea of extending a nationwide carbon tax.

Unpaid externalities

For a moment consider exactly what Big Oil is doing to our planet and how much money they're making in the process.The five biggest oil companies in the world — ExxonMobil, Chevron, Royal Dutch Shell, BP and ConocoPhillips — dump roughly a billion pounds of carbon into our air every single day. In that same day, they combine to make $342 million in profits. A modest carbon tax that would cut their daily profits by about $10 million and still total for them a staggering $332 million in profits each day is something they can afford as one part of a comprehensive energy policy. It's the bare minimum that carbon polluters should be paying considering the damages they're inflicting on our planet.

Pollution costs like global warming are called externalities. These externalities also include the medical costs of increased asthma and cancer care associated with a billion pounds of pollution being burned into the air above our communities every single day. They include the costs of ecological devastation caused when oil rigs explode in the Gulf of Mexico, or tankers run aground and spill open off the coast of Alaska. They also include the military costs of deploying armies around the world to ensure that the oil spigots in hostile regions keep flowing and the shipping lanes remain open. And they include the costs associated with climate change: costlier storms, rising sea levels, flooding and agricultural devastation. Carbon polluters themselves pay for few if any of these externalities. But instead they try to pass the bill along to all of us affected by their pollution. That needs to change to prevent disaster. When the day comes that businesses know pollution is at their expense, then they will reduce their carbon emissions.

Not a silver bullet

Capping carbon levels alone won't work. Research out of the MIT Global Change Institute[14] assumed a $20 tax per ton of carbon pollution, increasing 4% each year, and found that it would only modestly curb carbon emissions.

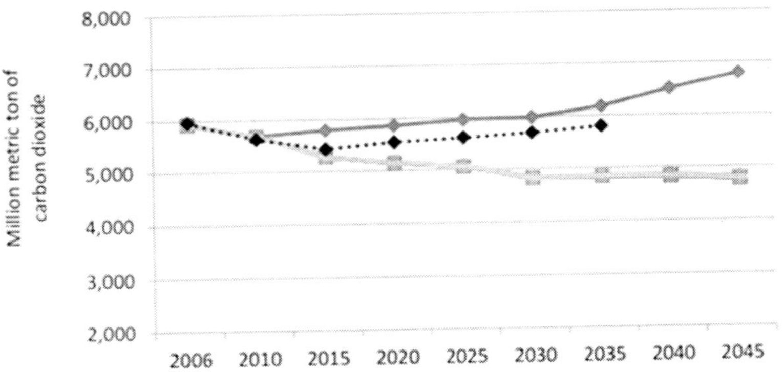

*Blue line: MIT reference case with no carbon tax. Black line: EIA reference. Green line: Scenario with MIT carbon tax in place.

This carbon tax scenario falls well short of even the White House's goals of cutting carbon emissions by 80% by 2050. And it would come nowhere close to staving off the projected temperature increases that could take our planet over the extinction tipping point.

So, a carbon tax needs to become part of a broader plan that not only eliminates incentives for the use of carbon-based fuels, but altogether phases them out and replaces them with new, clean-energy alternatives. This will require massive government investment in research and development.

Whether it was the Manhattan Project, which heralded in the nuclear age; the space race, bringing about the age of galactic exploration; or even Abraham Lincoln's expenditure of hundreds of billions (in today's dollars) to create a rail infrastructure across America, public investment has propelled every major industrial transition in our nation's history. It's time to make that investment again. This time to prevent extinction.

It's not economics; it's life and death

New research proves that the costs of doing nothing far outweigh the costs of a carbon tax or a green "Manhattan Project." A trio of scientists, Gail Whiteman of Erasmus University in the Netherlands and Chris Hope and Peter Wadhams of the University of Cambridge, released a study in July 2013 quantifying the consequences of a catastrophic methane release.

Their study notes: "As the amount of Arctic sea ice declines at an unprecedented rate, the thawing of offshore permafrost releases methane. A 50-gigatonne (Gt) reservoir of methane, stored in the form of hydrates, exists on the East Siberian Arctic Shelf. It is likely to be emitted as the seabed warms, either steadily over 50 years or suddenly. Higher methane concentrations in the atmosphere will accelerate global warming and hasten local changes in the Arctic, speeding up sea-ice retreat, reducing the reflection of solar energy and accelerating the melting of the Greenland ice sheet. The ramifications will be felt far from the poles."

The researchers show that with this sort of release of methane, it will increase global temperatures to that critical 2-degrees Celsius tipping point much faster than current projections. They suggest it will happen by 2035. And they calculate the economic consequences: "This will lead to an extra $60 trillion (net present value) of mean climate-change impacts for the scenario with no mitigation." To put that number into context, the annual GDP of the United States is around $15 trillion, and the annual GDP of the entire planet is around $65 trillion.

The economic cost of global climate change and particularly of a massive methane release are now well documented and clearly calculated. But the importance placed on these economic costs pales alongside the most dangerous threat of global warming: extinction.

Yes, we can and should project the economic costs of coastal cities washed away by rising sea levels or the $100 billion in costs associated with Hurricane Sandy and the drought in the Midwest in 2012. And that's just the United States — the rest of the world lost at least that much to global warming in 2012. But the monetary costs are of little importance when the very existence of the human species is threatened by the rapid warming of our planet.

Focusing on economic costs, though interesting and useful, shifts the debate in Washington away from the question of human survival and toward a question of fiscal policy. That waters down solutions and gives ammo to politicians who argue that while global warming may be occurring, it would harm our economy to force businesses to do anything about it. In reality, only arguing economics holds us back from immediately enacting serious curbs on carbon to save our planet.

Michael Mann told me, "We are already in the midst of one of the great extinction events of all-time in the entire history of the Earth. And climate change is one of the most serious contributing factors. We are literally changing the composition of the atmosphere and the climate at rates far greater than happened in the natural past."

"What that means," Mann added, "is we are subjecting living things to rates of change that they simply haven't evolved to deal with."

And that includes us.

Yes, there is a debate among scientists about whether this will be years, decades or centuries, but the question is, can we afford to take the chance when there is no plan B?

I believe we can't and I hope you do, too. Tag, you're it.

Endnotes

[1] Represents the percentage loss of marine life. Source: http://en.wikipedia.org/wiki/File:Extinction_intensity.svg

[2] Source: http://americablog.com/2013/05/global-warming-we-are-halfway-to-a-mass-extinction-event.html

[3] Robert prefers that his name not be capitalized

[4] http://www.ukmediacentre.pwc.com/News-Releases/-Business-as-usual-is-not-an-option-as-current-rates-of-emissions-reduction-point-to-6C-of-warming-PwC-Low-Carbon-Economy-Index-12e8.aspx

[5] http://climatechange.worldbank.org/sites/default/files/Turn_Down_the_Heat_Executive_Summary_English.pdf

[6] Source: http://www.washingtonpost.com/blogs/wonkblog/wp/2012/11/28/understanding-the-doha-climate-talks-in-three-charts/

[7] Source: http://www.nasa.gov/topics/earth/features/earth20130610.html

[8] Source: http://faculty.virginia.edu/setear/courses/globwarm/images.htm

[9] From research by Michael Mann. Source: http://www.globalwarming.org/2012/07/26/is-todays-climate-warmer-than-the-medieval-and-roman-warm-periods/

[10] Credit to Stefan Rahmstorf, grant Foster, Anny Cazenave. Source: http://iopscience.iop.org/1748-9326/7/4/044035/article

[11] Credit to Keynyn Brysse, Naomi Oreskes, Jessica O'Reilly, Michael Oppenheimer. Source: http://www.sciencedirect.com/science/article/pii/S0959378012001215

[12] Source: http://www.skepticalscience.com/print.php?n=779

[13] http://www.columbia.edu/~jeh1/2008/TargetCO2_20080407.pdf

[14] Source: http://www.washingtonpost.com/blogs/wonkblog/wp/2012/08/27/how-a-carbon-tax-could-help-the-u-s-avert-the-fiscal-cliff/

Acknowledgments

This book would not be where it is without the help of writer/researcher/editor Sam Sacks. Sam, thanks so much for your help!

Others who have helped make this book what it is include Earl Katz, Louise Hartmann, Shawn Taylor, Kenneth Kales, Bill Gladstone, George and Leonardo DiCaprio, Leila Connors, James Cameron and Justin Hartmann. Also, special thanks to RT America and Misha Solodovnikov for help with graphics.

The many scientists who did so much work in this field before we arrived to chronicle it deserve so much credit. We are truly "standing on the shoulders of giants." In particular, Michael Mann deserves a special thanks both for helping us and helping to awaken the world with his "hockey stick" formulation.

THE CRASH OF 2016: The Plot to Destroy America—and What We Can Do to Stop It

By Thom Hartmann

On Sale Date: November 12, 2013

Hardcover: ISBN: 978-0-446-58483-8 • $28.00 • 320 pages

E-Book ISBN: 978-0-446-58481-4 • $14.99

www.twelvebooks.com

Amazon: http://www.amazon.com/The-Crash-2016-Destroy-America/dp/0446584835/ref=tmm_hrd_title_0

Amazon Kindle: http://www.amazon.com/The-Crash-2016-Destroy-America/dp/0446584835

B&N: http://www.barnesandnoble.com/w/the-crash-of-2016-thom-hartmann/1115779224?ean=9780446584814

Apple: https://itunes.apple.com/us/book/the-crash-of-2016/id599675466?mt=11

From the Crash of 2016

The United States is more vulnerable today than ever before—including during the Great Depression and the Civil War—because the pillars of democracy that once supported a booming middle class have been corrupted. Without them, America teeters on the verge of the next Great Crash.

The United States is in the midst of an economic implosion that could make the Great Depression look like child's play. In THE CRASH OF 2016, Thom Hartmann argues that the United States will soon disintegrate to reveal the rotting core where corporate and billionaire power and greed have replaced democratic infrastructure and governance. Our once-enlightened political and economic systems have been manipulated to ensure the success of only a fraction of the population at the expense of the rest of us.

The result is a "for the rich, by the rich" scheme leading to policies that benefit only the highest bidders. Hartmann outlines the destructive forces—planted by Lewis Powell in 1971 and brought to fruition with the "Reagan Revolution"—that have looted our nation during the past decade, and how their actions fit into a cycle of American history that lets such forces rise to power every four generations.

However, a backlash is now palpable against the "Economic Royalists"—a term coined by FDR to describe those hoarding power and wealth—including the banksters, oligarchs, and politicians who have plunged our nation into economic chaos and social instability.